Guided Notebook
Beginning Statistics
+Integrated Review

A division of Quant Systems, Inc.

546 Long Point Road
Mount Pleasant, SC 29464

Copyright © 2019, 2017, 2016 by Hawkes Learning / Quant Systems, Inc. All rights reserved.

No part of this publication may be reproduced, stored in a retrieval system, or transmitted in any form or by any means, electronic, mechanical, photocopying, recording, or otherwise, without the prior written consent of the publisher.

Printed in the United States of America

10 9 8 7 6 5 4 3 2

ISBN: 978-1-944894-83-2

TABLE OF CONTENTS

Chapter 1.R
Introduction to Statistics
Integrated Review

1.R.1	Problem Solving with Whole Numbers	6
1.R.2	Introduction to Decimal Numbers	7

Chapter 2.R
Graphical Descriptions of Data
Integrated Review

2.R.1	Introduction to Fractions and Mixed Numbers	10
2.R.2	Decimals and Fractions	12
2.R.3	Decimals and Percents	13
2.R.4	Reading Graphs	14
2.R.5	Constructing Graphs from a Database	15
2.R.6	The Real Number Line and Inequalities	16

Chapter 3.R
Numerical Descriptions of Data
Integrated Review

3.R.1	Addition with Real Numbers	20
3.R.2	Subtractions with Real Numbers	21
3.R.3	Multiplication and Division with Real Numbers	22
3.R.4	Exponents and Order of Operations	24
3.R.5	Evaluating Algebraic Expressions	25
3.R.6	Evaluating Radicals	26

Chapter 4.R
Probability, Randomness, and Uncertainty
Integrated Review

4.R.1	Multiplication and Division with Fractions and Mixed Numbers	30
4.R.2	Least Common Multiple (LCM)	31
4.R.3	Addition and Subtraction with Fractions	32
4.R.4	Fractions and Percents	33

Chapter 5.R

Discrete Probability Distributions
Integrated Review

5.R.1 Order of Operations with Real Numbers 36
5.R.2 Solving Linear Inequalities 37

Chapter 6.R

Normal Probability Distributions
Integrated Review

6.R.1 Area 42
6.R.2 Solving Linear Equations: $ax + b = c$ 43
6.R.3 Working with Formulas 44

Chapter 7.R

The Central Limit Theorem
Integrated Review

7.R.1 Ratios and Ratess 46

Chapter 10.R

Hypothesis Testing
Integrated Review

10.R.1 Translating English Phrases and Algebraic Expressions 48
10.R.2 Order of Operations with Fractions and Mixed Numbers 49

Chapter 12.R

Regression, Inference, and Model Building
Integrated Review

12.R.1 The Cartesian Coordinate System 52
12.R.2 Graphing Linear Equations in Two Variables: $Ax + By = C$ 54
12.R.3 The Slope-Intercept Form: $y = mx + b$ 56

Chapter 1.R
Introduction to Statistics
Integrated Review

Problem Solving with Whole Numbers

- Basic Strategy for Solving Word Problems:

    1. _____ the problem carefully.

    2. _____ any type of figure or _____ that might be helpful and decide what operations are needed.

    3. Perform the _____ to solve the problem.

    4. _____ your work.

- Key words when solving word problems are:

Addition	Subtraction	Multiplication	Division

- To find the average of a set of numbers, you will need to:

    1. Find the _____ of the given set of numbers.

    2. _____ this sum by the total number of numbers in the set. This quotient is called the _____ of the given set of numbers.

Introduction to Decimal Numbers

o Decimal notation uses a _____ _____ system and a _____ point, with whole numbers written to the _____ and fractions written to the _____ of the decimal point.

o To read or write a decimal number:

   o _____ (or write) the whole number.

   o Read (or write) the word "_____" in place of the decimal point.

   o Read (or write) the _____ part as a whole number. Then name the fraction with the name of the last _____ on the right. Add "th" to the end of the fraction place value.

   o Remember that if there is not a whole number, you can put a _____ to the left of the decimal point.

o Write the following mixed number as a decimal number:

   $3\frac{4}{100} = $ _____ . _____ _____

   and it would be read as _____ AND _____ _____

o When writing seventeen and 5 thousandths in decimal notation you would have _____ holders in the tenths and hundredths place values.

It would look like 17.005.

o When comparing decimals:

   o Moving _____ to _____, compare digits with the same _____ value.

   o When one compared digit is larger, the _____ is larger.

8   1.R.2   Introduction to Decimal Numbers

- Compare the following values:

    5.789   Notice that the numbers are lined up, for easier comparison.
    5.754

    When moving from left to right, the digits are the same until the _____ place values. Those are going to be used to compare. Since the 8 is a larger value than 5, _____ is a larger value.

- Rules for rounding decimals:

    - Look at the digit to the _____ of the place of desired accuracy.

    - If this digit is 5 or _____, make the digit in the desired place of accuracy one larger and replace all digits to the right with zeros.

    All digits to the left remain unchanged unless a 9 is made one larger. This effectively changes the 9 to 10 which means the next digit to the left must be increased by 1.

    - If this digit is _____ than 5, leave the digit in the desired place of accuracy as it is and replace all digits to the right with zeros.

    All digits to the left remain unchanged.

    - Zeros **to the right of the place of accuracy** and to the right of the decimal point must be _____. In this way the place of accuracy is clearly understood. If a rounded number has a 0 in the desired place of accuracy, then that 0 remains.

Round 13.2687 to the nearest hundredth.

   The digit in the hundredths is _____ .

   The digit in the place value to the right of the hundredths is _____ .

   Since that digit is greater than 5, the 6 in the hundredths place value changes to a 7.

   The rounded value is _____._____ .

Chapter 2.R
Graphical Descriptions of Data
Integrated Review

Introduction to Fractions and Mixed Numbers

- The parts of a fraction are (shown below):

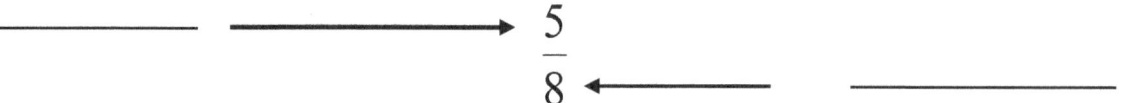

- Fractions are used to indicate _____ of a whole.

- Use the picture below to write a fraction representing the shaded portion of the shape:

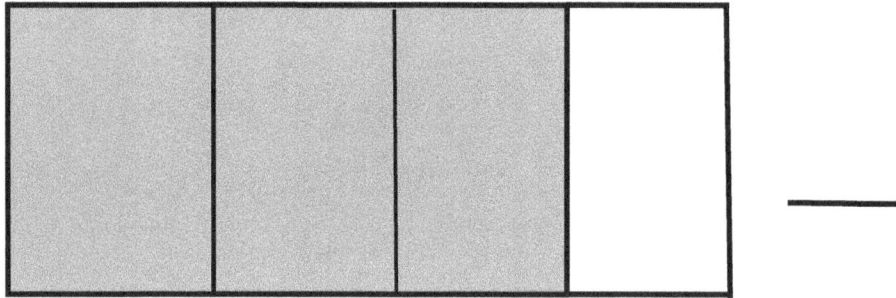

- The fraction $\dfrac{2}{7}$ represents _____ of _____ equal parts.

- Whole numbers can be thought of as fractions with a denominator of

  _____.

- Fraction notation represents the operation of _____.

- There are two rules to keep in mind when working with the value of zero in a fraction:

- For any nonzero value, b, $\dfrac{0}{b} = 0$. An example of this would be $\dfrac{0}{4} = 0$.

- For any value of a, $\dfrac{a}{0} =$ undefined. An example of this would be $\dfrac{6}{0} =$ undefined.

- Steps to multiply fractions:

    1. Multiply the _____.

    2. Multiply the _____.

2.R.1 Introduction to Fractions and Mixed Numbers

- **Multiply the following fraction:** $\dfrac{2}{3} \cdot \dfrac{1}{5} =$ _____ = _____

    o To find an equivalent fraction, you need to _____ the numerator and the denominator by the _____ nonzero whole number.

    o Find an equivalent fraction for $\dfrac{2}{3} =$ _____ by multiplying both the numerator and the denominator by 5.

    o There are two steps to reducing fractions to lowest terms:

    1. Factor the _____ and _____ into prime factors.

    2. Use the fact that $\dfrac{k}{k} = 1$ and divide out all of the _____ factors.

    o What is the common factor that can be divided out of the fraction, $\dfrac{4}{12}$?

- **What is $\dfrac{4}{12}$ in lowest terms?**

    o A mixed number is the sum of a _____ and a _____ fraction.

    o To change a mixed number to an improper fraction, you need to:

    1. Multiply the whole number by the _____ of the proper fraction.

    2. Add the _____ of the proper fraction to this product.

    3. Write this _____ over the denominator of the fraction.

    o Change $3\dfrac{1}{2}$ to an improper fraction:

    Multiply the whole number by the denominator: ____ · ____ = ____

- o Add the numerator to the product from above: ____ + ____ = ____

    o Write this sum over the denominator: ____

Decimals and Fractions

- To change from decimal numbers to fractions:

  - A terminating decimal number can be written in fraction form by writing a fraction with the following:

    - a _____ that consists of the whole number formed by all the digits of the decimal number

    - a _____ that is the power of ten that names the position of the last digit on the right

- To change from a fraction to a decimal number:

  - To change a fraction to decimal form, we divide the numerator by the denominator.

    - If the remainder is eventually 0, the decimal number is said to be _____.

    - If the remainder is never 0, the decimal number is said to be _____.

- Nonterminating decimal numbers can be _____ or nonrepeating.

- A _____ repeating decimal (also called an infinite repeating decimal number) has a repeating pattern to its digits.

- Every fraction with a whole number numerator and nonzero denominator is either terminating or repeating. Such numbers are called _____ numbers.

- Sometimes, changing fractions to decimal form may involve _____ the decimal form of a number and settling for an approximate answer. To have a more accurate answer, we may need to change the number from decimal form to _____ form and then perform the operations.

Decimal and Percents

- The word percent comes from the Latin *per centum*, meaning per _____.

- Percent means _____, or the ratio of a number to 100.

- The symbol _____ is called the percent sign. This sign has the same meaning as the _____ $\frac{1}{100}$.

- Changing fractions with denominators of 100 to percents:

- Example: $\frac{25}{100} = 25\%$

  The _____ did not change.

- Example: $\frac{3.8}{100} = 3.8\%$

  The _____ is not changed, the decimal point doesn't move if the _____ is 100.

- To change a decimal to a percent:

1. Move the decimal point two places to the _____.

2. Write the _____ sign.

    - Example: $0.56 = 56\%$
    - Example: $0.345 = 34.5\%$
    - Example: $0.02 = 2\%$

- To change percents to a decimal number:

1. Move the decimal two places to the _____.

2. Delete the _____ sign.

    - Example: $97\% = 0.97$
    - Example: $68.5\% = 0.685$
    - Example: $0.64\% = 0.0064$

2.R.4 Reading Graphs

Reading Graphs

- The Purposes of Four Types of Graphs
    - Bar Graphs: to emphasize _____ amounts.
    - Circle Graphs: to help in understanding _____ or parts of a _____. Circle graphs are also called _____ charts.
    - Line Graphs: to indicate _____ or _____ over a period of time.
    - Histograms: to indicate data in _____ (a _____ or interval of numbers).
- A common characteristic of all graphs is that they are intended to _____ information about numerical data _____ and _____.
- All graphs should:
    - Be clearly _____.
    - Be easy to _____.
    - Have appropriate _____.
- Keep these terms in mind when working with histograms:
    - Class: an _____ (or range) of numbers that contains data items.
    - Lower class limit: the _____ whole number that belongs to a class.
    - Upper class limit: the _____ whole number that belongs to a class.
    - Class boundaries: numbers that are halfway between the _____ limit of one class and the _____ limit of the next class.
    - Class width: the _____ between the class boundaries of a class (the width of each bar).
    - Frequency: the _____ of data items in a class.

Constructing Graphs from a Database

o A bar graph may have either _____ or _____ bars. In both cases, the length of each bar represents the _____ of the data in a category being graphed.

o When constructing a vertical bar graph, it is important to remember that the scale must be _____.

o A circle graph is a circle that is marked in sectors (pie-shaped wedges) that correspond to _____ of data in each category represented.

o The sum of all the central angles of the sectors in a circle graph will equal _____.

The Real Number Line and Inequalities

- A number line is a picture of different types of _____ and their relationships to each other.

- The graph of a number is the point that _____ to the number and the number is called the _____ of the point.

- On a horizontal number line, the point one unit to the left of 0 is the _____ of 1. It is called negative 1 and is symbolized −1.

- The negative sign (−) indicates the _____ of a number as well as a _____ number. It is also used to indicate _____.

- The set of numbers consisting of the _____ numbers and their _____ is called the set of integers: $\mathbb{Z} = \{..., -3, -2, -1, 0, 1, 2, 3, ...\}$.

- The _____ numbers are also called _____ integers. Their opposites are called _____ integers. Zero is its _____ opposite and is neither positive nor negative (0 = −0).

- The opposite of a positive integer is a _____ integer, and the opposite of a negative integer is a _____ integer.

- A rational number is a number that can be written in the form of $\frac{a}{b}$, where a and b are _____ and $b \neq$ _____. ($\neq$ is read "is not equal to").

- A rational number is a number that can be written in _____ form as a _____ decimal or as an infinite _____ decimal.

- Irrational numbers can be written as infinite _____ decimal numbers.

- All rational numbers and irrational numbers are classified as _____ numbers (ℝ) and can be written in some decimal form.

- Symbols of equalities and inequalities:

    - = is _____ to

    - ≠ is _____ equal to

    - < is _____ than

    - \> is _____ than

    - ≤ is less than or _____ to

    - ≥ is greater than or _____ to

2.R.6 The Real Number Line and Inequalities

Chapter 3.R
Numerical Descriptions of Data
Integrated Review

Addition with Real Numbers

- The sum of two positive real numbers is _____.

- The sum of two negative real numbers is _____.

- The sum of a positive real number and a negative real number may be _____ or _____ (or _____) depending on which number is _____ from 0.

- To add two real numbers with like signs:

1. Add their _____ _____

2. Use the _____ sign

- To add two real numbers with unlike signs:

1. _____ their absolute values (the smaller from the larger)

2. Use the sign of the number with the _____ absolute value

- The _____ sign may be omitted when writing _____ numbers, but the _____ sign must always be written for _____ numbers.

- If there is no sign in front of a real number, the real number is understood to be _____.

- A number is said to be a _____ or to _____ an equation if it gives a true statement when substituted for the variable.

Subtraction with Real Numbers

- The opposite of a real number is called its _____ _____.

- The sum of a number and its additive inverse is _____.

- Adding positive numbers "_____ ____" or "_____" the numbers in a _____ direction (moving _____ on the number line).

- Adding negative numbers "_____ ____" or "_____" the numbers in a _____ direction (moving _____ on the number line).

- In subtraction, you want to find the "_____" two numbers.

- To subtract, _____ the _____ of the number being subtracted.

- To find the change in value between two numbers, take the end value and _____ the beginning value.

- This process looks like:

    Change in value = (_____ value) − (_____ value)

Multiplication and Division with Real Numbers

- If *a* and *b* are positive real numbers, then:

    - The product of two positive numbers is _____: $a \cdot b = +ab$.

        Example: $8(14) = 122$

    - The product of two negative numbers is _____: $(-a)(-b) = +ab$.

        Example: $-4(-5) = 20$

    - The product of a positive number and a negative number is _____: $a(-b) = -ab$.

        Example: $10(-3) = -30$

    - The product of 0 and any number is _____: $a \cdot 0 = 0$ and $(-a) \cdot 0 = 0$.

        Example: $71(0) = 0$

- If *a* and *b* are positive real numbers (where $b \neq 0$), then:

    - The quotient of two positive numbers is _____: $\dfrac{a}{b} = +\dfrac{a}{b}$

        Example: $\dfrac{14}{2} = 7$

    - The quotient of two negative numbers is _____: $\dfrac{-a}{-b} = +\dfrac{a}{b}$

        Example: $\dfrac{-18}{-6} = 3$

    - The quotient of a positive number and a negative number is _____:

        $\dfrac{-a}{b} = -\dfrac{a}{b}$ and $\dfrac{a}{-b} = -\dfrac{a}{b}$

        Example: $\dfrac{-144}{12} = -12$

        Example: $\dfrac{88}{-11} = -8$

- Quick Reference Facts with Multiplication and Division with Real Numbers:

    - If the numbers have the _____ sign, both the product and quotient will be _____.

    - If the numbers have _____ signs, both the product and quotient will be _____.

Exponents and Order of Operations

- When looking at 5^3, you have the following parts:

    - The base is: _____

    - The exponent is: _____

    - The product is: _____

    - The exponential expression is: _____

- In expressions with exponent 2, the base is said to be _____.

- In expressions with exponent 3, the base is said to be _____.

- With other exponents, the base is said to be "_____."

- For any number, a, $a^1 = a$. An example would be $7^1 =$ _____.

- For any nonzero number, a, $a^0 = 1$. An example would be $7^0 =$ _____.

- The rules for order of operations are:

    1. Simplify within _____ _____, such as parentheses ( ), brackets [ ], or braces { }. Start with the _____ group.

    2. Evaluate any numbers or expressions indicated by _____.

    3. Moving from _____ to _____, perform any _____ or _____ in the order in which it appears.

    4. Moving from _____ to _____, perform any _____ or _____ in the order in which it appears.

Evaluating Algebraic Expressions

- The expression $-x^2$ will have a _____ value.

    - An example of this is $-6^2 =$ _____ .

- The expression $(-x)^2$ will have a _____ value.

    - An example of this is $(-6)^2 =$ _____ .

- To evaluate an algebraic expression, you will need to:

    1.

    2.

    3.

Evaluating Radicals

o Complete the tables of Squares and Perfect Squares below:

Integers (n)	1	2	3	4	5	6	7	8	9	10
Perfect Squares (n^2)										

Integers (n)	11	12	13	14	15	16	17	18	19	20
Perfect Squares (n^2)										

o Finding the square root of a number is the opposite of _____ a number.

o The symbol _____ is called a radical sign.

o The number under the radical sign is called the _____.

o The complete expression, such as _____ , is called a radical or radical expression.

o If a is a nonnegative real number, then _____ is the principal square root of a.

o If a is a nonnegative real number, _____ is the negative square root of a.

o A number is cubed when it is used as a factor 3 times.

o Complete the cubes of numbers below:

$1^3 =$	$2^3 =$	$3^3 =$	$4^3 =$	$5^3 =$
$6^3 =$	$7^3 =$	$8^3 =$	$9^3 =$	$10^3 =$

o If a is a real number, then _____ is the cube root of a.

o In the cube root expression $\sqrt[3]{a}$, the number 3 is called the _____.

o In a square root expression such as $\sqrt{a}$, the index is understood to be 2 and is _____ written.

o Expressions with square roots and cube roots, as well as other roots, are called _____.

3.R.6 Evaluating Radicals

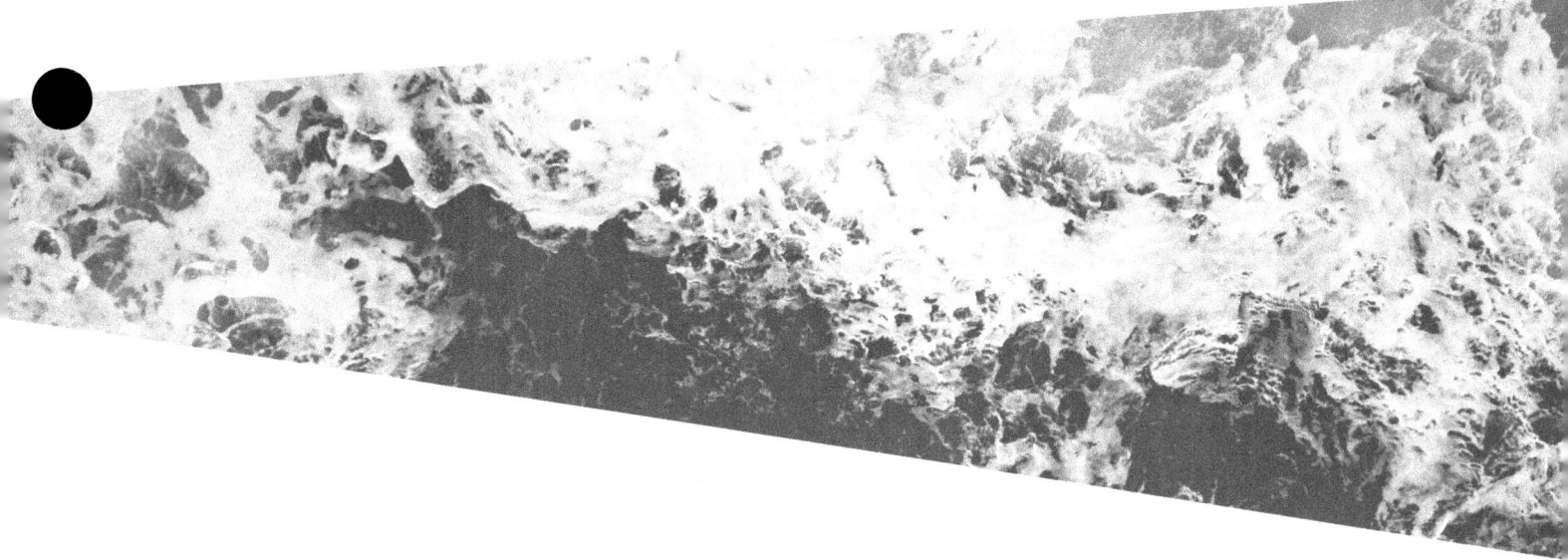

Chapter 4.R
Probability, Randomness, and Uncertainty
Integrated Review

Multiplication and Division with Fractions and Mixed Numbers

- To multiply with mixed numbers, you will need to change the mixed numbers to _____ _____ and then multiply the fractions.

- You can multiply fractions and mixed numbers while reducing at the same time by using _____ factors.

- Multiply and reduce to lowest terms:

$$1\frac{2}{5} \cdot \frac{5}{8} = \frac{ \cdot }{ \cdot \cdot } = \frac{}{}$$

- Remember that if all of the factors in the _____ or the _____ divide out, then the value of ____ must be used as a factor.

- The reciprocal of $\frac{a}{b}$ is $\frac{b}{a}$ (when a and b do not equal 0).
  The product of a nonzero number and its reciprocal is always ____.

- Match the following values and its reciprocal:

15	2
$\frac{7}{8}$	$\frac{1}{15}$
$\frac{1}{2}$	$\frac{8}{7}$

- To divide fractions by any nonzero number, you will _____ by its reciprocal.

  For example, $\frac{1}{2} \div \frac{2}{5} = \frac{1}{2} \cdot \frac{}{}$.

Least Common Multiple (LCM)

- The _____ of a number are the products of that number with the counting numbers.

- The least common multiple (LCM) of two (or more) whole numbers is the _____ number that is a multiple of each of these numbers.

- Steps to find the LCM of Counting Numbers:

    1. Find the _____ _____ of each number.

    2. Identify the prime factors that appear in _____ one of the prime factorizations.

    3. Form the product of these primes using each prime the _____ number of times it appears in any one of the prime factorizations.

- Find the LCM of 24 and 36.

    1. The prime factorization for 24 is: _____ .

       The prime factorization for 36 is: _____ .

    2. You will use the factors of _____ .

    3. Multiply the factors together to get the LCM of _____ .

Addition and Subtraction with Fractions

- To **add** fractions with the **same denominator**, you will need to:

    - _____ the numerators.

    - Keep the _____ denominator.

    - _____, if possible.

- To **add** fractions with **different denominators**, you will need to:

    - Find the least common _____ (LCD). Remember that the LCD is the least common multiple for the denominators.

    - Change each fraction into an _____ fraction with that denominator.

    - _____ the new fractions.

    - Reduce, if possible.

- A common error that occurs when adding fractions that needs to be avoided is _____ out across the addition sign.

- To **subtract** fractions with the **same denominator**, you will need to:

    - _____ the numerators.

    - Keep the _____ denominator.

    - Reduce, if possible.

- To **subtract** fractions with **different denominators**, you will need to:

    - Find the least common _____ (LCD).

    - Change each fraction into an _____ fraction with that denominator.

    - _____ the new fractions.

    - Reduce, if possible.

Fractions and Percents

- If a fraction has denominator _____, it can be changed to a percent by writing the _____ and adding the _____ sign.

- If the denominator is a factor of _____ (2, 4, 5, 10, 20, 25, or 50), the fraction can be changed to an equivalent fraction with denominator of _____ and then changed to a percent.

- When fractions do NOT have factors of 100 as denominators you will need to:

- Change the fraction to decimal form by _____, either by long division or with a calculator (depending on instructions from instructor).

- Move the decimal two places to the _____ and write the _____ symbol.

  Example: $\dfrac{3}{4} = 4\overline{)3} = 0.75 = 75\%$

- Helpful calculation tips:

    - The numerator goes _____ the division symbol and the denominator goes _____ of the division symbol.

    - If you are using a calculator, you will type the numerator _____, then the division symbol, and then the _____, followed by enter/equal.

        - To change a percent to a fraction or a mixed number:

    - Write the percent as a fraction with 100 as the _____ and drop the _____ symbol.

    - _____ the fraction, if possible.

      Example: $80\% = \dfrac{80}{100} = \dfrac{2 \cdot 2 \cdot 2 \cdot 2 \cdot 5}{2 \cdot 2 \cdot 5 \cdot 5} = \dfrac{4}{5}$

Chapter 5.R
Discrete Probability Distributions
Integrated Review

Order of Operations with Real Numbers

- Order of Operations:

    - Simplify within grouping symbols, such as

        _____ ( ), _____ [ ], and

        _____ { }, working from the innermost grouping

        _____.

    - Find any powers indicated by _____.

    - Moving from _____ to _____, perform any

        _____ or _____ in the order they appear.

    - Moving from _____ to _____, perform any

        _____ or _____ in the order they appear.

- Keep in mind that there are other _____ symbols like the

    _____ _____ bars (such as |3+5|), the

    _____ _____ (as in $\frac{14+7}{3}$), and the

    _____ _____ symbol (such as $\sqrt{40+9}$).

- Even though the mnemonic PEMDAS is helpful, remember that multiplication

    and division are performed as they _____, left to right.

- Also, addition and subtraction are performed as they _____, left to right.

Solving Linear Inequalities

- A _____ is a collection of objects or numbers.

- The items in the set are called _____.

- Sets are indicated with _____, that look like _____.

- If elements are listed within the braces, the set is said to be in _____ form.

- The symbol ∈ is read "is an _____ of" and is used to indicate that a particular number belongs to a _____.

- If the elements in a set can be _____ the set is said to be _____.

- If the elements cannot be counted the set is said to be _____.

- If a set has absolutely no elements, it is called the _____ set or null set and is written in the form { } or with the special symbol _____.

- The notation _____ is read "the set of all x such that …" and is called _____-builder notation.

- The vertical bar that looks like _____ is read "such that."

- Example: $\{x \mid x \text{ is an even integer}\}$ is read "the set of all x _____ that x is an even integer."

- The _____ (symbolized ∪, as in A ∪ B) of two (or more) sets is the set of all elements that belong to either _____ set or the _____ set or to _____ sets.

- The _____ (symbolized ∩, as in A ∩ B) of two (or more) sets is the set of all elements that belong to _____ sets.

5.R.2 Solving Linear Inequalities

- The word _____ is used to indicate union and the word _____ is used to indicate intersection.

- Suppose that a and b are two real numbers and that $a<b$. The set of all real numbers between a and b is called an _____ of real numbers.

- A _____ interval such as $-3<x<6$ is read "x is greater than -3 and x is less than 6."

Complete the table below based on the information in Learn:

Type of Interval	Algebraic Notation	Interval Notation	Graph

- In an open interval, _____ endpoint is included.

- In a closed interval, _____ endpoints are included.

- In a half-open interval, only _____ endpoint is included.

- The symbol for infinity, which looks like _____ or _____, is not a number. It is used to indicate that the interval is to include _____ real numbers from some point on (either in the positive direction or the negative direction) without _____.

- When solving linear inequalities, _____ or _____ both sides of an inequality by a negative number causes the "sense" of the inequality to be _____.

- Rules for solving linear inequalities:

    1. Simplify each side of the inequality by removing any _____ symbols and _____ like terms.

    2. Use the _____ property of equality to add the opposites of constants or variable expressions so that variable expressions are on one side of the inequality and constants are on the other.

    3. Use the _____ property of equality to multiply both sides by the reciprocal of the coefficient of the variable (that is, divide both sides by the coefficient) so that the new coefficient is 1. If this coefficient is negative, reverse the sense of the inequality.

    4. A quick (and generally satisfactory) check is to select any one number in your solution and _____ it into the original inequality.

- Compound inequalities have _____ parts and can arise when a variable or variable expression is to be between two numbers.

Chapter 6.R
Normal Probability Distributions
Integrated Review

Area

- Area is a measure of the _____ of (or surface enclosed by) a plane figure and is measured in _____ units.

Units of Area

Basic U.S. Units of Area	Metric Units of Area
Square _____ (in.²)	Square _____ (mm²)
Square _____ (ft²)	Square _____ (cm²)
Square _____ (yd²)	Square _____ (km²)
Square _____ (mi²)	Square _____ (m²)

- Be sure to label your answers with the _____ units.

- In triangles and other figures, we have used the letter _____ to represent the height of the figure. The height is also called the _____ and is perpendicular to the base.

- Common area formulas:

_____ $A = \dfrac{bh}{2}$

_____ $A = l \cdot w$

_____ $A = s^2$

_____ $A = b \cdot h$

_____ $A = \dfrac{h(b+c)}{2}$

Solving Linear Equations: $ax + b = c$

- The general procedure for solving linear equations is now a _____ of the procedures stated in another section.

- Steps for solving equations in the $ax+b=c$ format.

    1. _____ like terms on both sides of the equation.

    2. Use the _____ principle of equality and add the opposite of the constant b to both sides.

    3. Use the _____ (or division) principle of equality to multiply both sides by the reciprocal of the coefficient of the variable (or divide both sides by the coefficient itself). The coefficient of the variable will become +1.

    4. Check your answer by _____ it for the variable in the original equation.

- Keep in mind that checking can be quite _____-_____ and need not be done for every problem. This is particularly important on _____. You should check only if you have time after the entire exam is _____.

    Example:

    $3x - 8 = 28$ Add 8 to each side.

    $3x = 36$ Divide both sides by 3.

    $x = 12$

44   6.R.3   Working with Formulas

Working with Formulas

o Formulas are general _____ or principles stated _____.

Below is a table of formulas and their meanings. You will use these in this lesson.

Formula	Meaning
$I = Prt$	The simple interest I earned by investing money is equal to the product of the _____ P times the rate of interest r times the time t in one year or part of a year.
$C = \dfrac{5}{9}(F - 32)$	Temperature in degrees Celsius C equals $\dfrac{5}{9}$ times the _____ between the Fahrenheit temperature F and _____.
$d = rt$	The _____ traveled d equals the product of the rate of _____ r and the time t.
$P = a + b + c$	The perimeter P of a triangle is equal to the sum of the _____ of the three sides a, b, and c.
$F = kAv^2$	The lifting _____ F is equal to the product of the constant k by the area A and by the square of the plane's velocity v.

Chapter 7.R
The Central Limit Theorem
Integrated Review

Ratios and Rates

- Two meanings for fractions are:

    - To indicate _____ of a whole.

    - To indicate _____.

- A ratio is a _____ of two quantities by

    _____.

- A ratio, 3 days to 4 nights, can be written in three ways:

    1.

    2.

    3.

- Ratios have the following characteristics:

    - Ratios can be _____.

    - Whenever units of the numbers in a ratio are the same, the ratio has

        _____ units, otherwise known as an _____ number.

    - When the numbers in a ratio have different units, then the numbers must be

        _____ to clarify what is being compared.

- A _____ is a statement that two ratios are equal.

- A proportion is _____ if the cross products are equal.

Chapter 10.R
Hypothesis Testing
Integrated Review

Translating English Phrases and Algebraic Expressions

Addition	Subtraction	Multiplication	Division	Exponent (Powers)
Add	Subtract (from)	Multiply	Divide	Square of
Sum	Difference	Product	Quotient	Cube of
Plus	Minus	Times		
More than	Less than	Twice		
Increased by	Decreased by	Of (with fractions and percents)		
	less			

Key Words for Translating Phrases

- Remember that in each case "a number" or "the number" implies the use of a _____ (an unknown quantity).

- Be very careful when writing and/or interpreting expressions indicating _____. Be sure that the subtraction is in the order indicated by the wording in the problem. The same is true with expressions involving _____.

- _____ and _____ are done with the values in the same order that they are given in the problem.

Order of Operations with Fractions and Mixed Numbers

- To compare two fractions (to find which one is larger or smaller):

    - Find the least common _____.

    - Change each fraction to an _____ fraction with that _____.

    - Compare the _____.

- Rules for order of operations:

    - Simplify within _____ symbols, like parentheses, brackets, and braces, starting with the _____ most group.

    - Evaluate any numbers or expressions indicated by _____.

    - Moving from _____ to _____, perform any multiplication or division in the order in which they appear.

    - Moving from _____ to _____, perform any addition or subtraction in the order in which they appear.

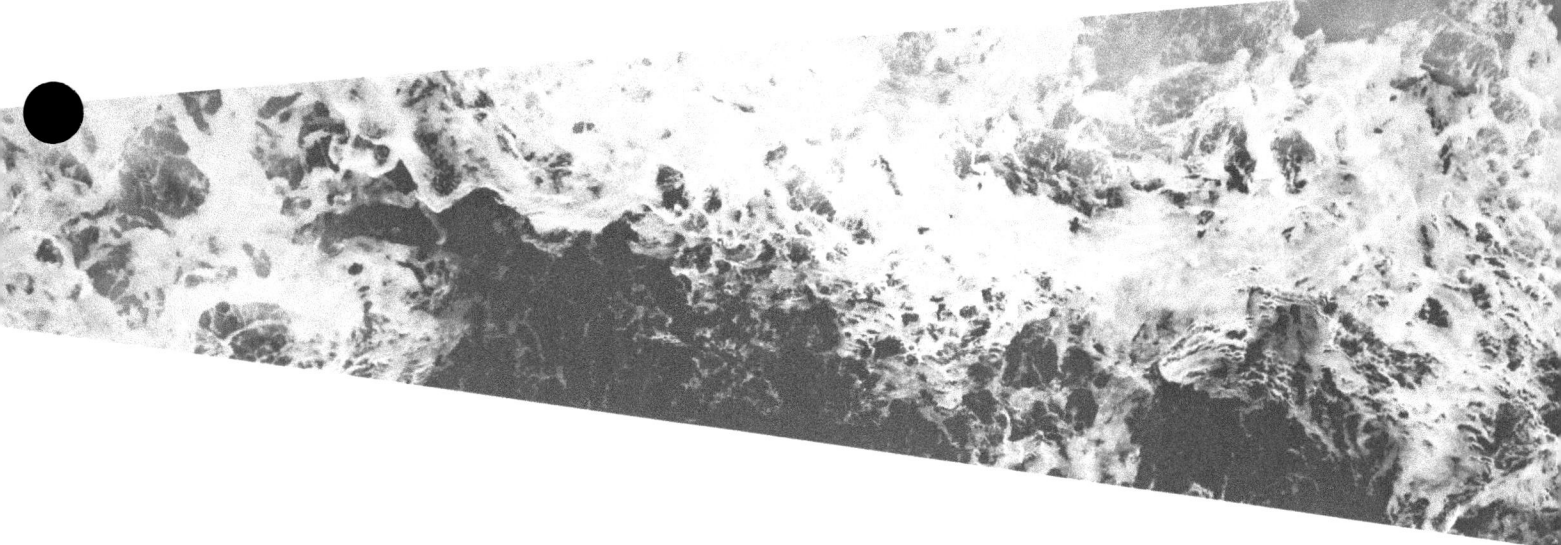

Chapter 12.R
Regression, Inference, and Model Building
Integrated Review

The Cartesian Coordinate System

- The Cartesian coordinate system (or the rectangular coordinate system) is based on a relationship between _____ in a plane and ordered pairs of real numbers.

- For an equation written like $y = 5x + 7$, ordered pairs are written in the form of (___, _____).

- In an ordered pair, x is called the first _____ and y is called the _____ coordinate.

- To find ordered pairs that satisfy an equation in two variables, we can choose any value for one variable and find the corresponding value for the other variable by _____ into the equation.

- There is an _____ number of such ordered pairs. Any real number could have been chosen for x and the corresponding value for y calculated.

- In an ordered pair of the form (x,y), the first coordinate x is called the _____ variable and the second coordinate y is called the _____ variable.

- Label the following information on the coordinate plane below: x-axis, y-axis, quadrants (4), and origin.

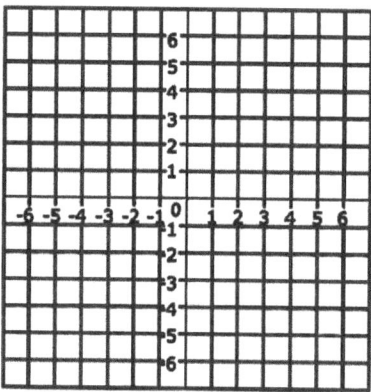

- In quadrant I, the x is _____ and the y is _____.

- In quadrant II, the *x* is _____ and the *y* is _____.

- In quadrant III, the *x* is _____ and the *y* is _____.

- In quadrant IV, the *x* is _____ and the *y* is _____.

- There is a one-to-one _____ between points in a plane and ordered pairs of real numbers.

- Unless otherwise stated, assume that the grid lines are _____ unit apart.

- To plot an *x*-coordinate, you will move_____ for negative values and _____ for positive values.

- To plot any *y*-coordinate, you will move_____ for negative values and up for _____ values.

Graphing Linear Equations in Two Variables: $Ax + By = C$

- To find some solutions for an equation like $2 - 2x = y$, you will need to:

    1. choose arbitrary values for _____

    2. find the corresponding values for _____ by substituting into the equation.

- The standard form of an equation is _____ where A, B, and C are real numbers and A and B are not both equal to _____.

- In the standard form _____, A and B may be positive, negative, or 0, but A and B cannot both equal 0.

- Every line corresponds to some _____ _____, and the graph of every linear equation is a line.

- Two points determine a _____.

- Steps to graph a linear equation in two variables:

    1. Locate any _____ points that satisfy the equation. (Choose values for x and y that lead to _____ solutions. Remember that there is an _____ number of choices for either x or y. But, once a value for x or y is chosen, the corresponding value for the other variable is found by substituting into the equation.)

    2. Plot these _____ points on a Cartesian coordinate system.

    3. Draw a _____ through these two points. (Note: Every point on that line will satisfy the equation.)

4. To check: Locate a _____ point that satisfies the equation and check to see that it does indeed lie on the line.

- By letting $x =$ _____, you will locate the point on the graph where the line crosses (or intercepts) the y-axis. This point is called the y-_____ and is of the form (___,y).

- The x-intercept is the point found by letting $y =$ _____. This is the point where the line crosses (or intercepts) the x-axis and is of the form (x,_____).

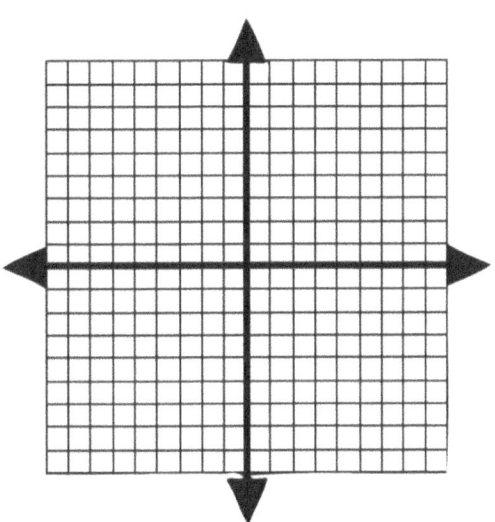

- When the intercepts result in a point with fractional (or decimal) coordinates and _____ is involved, then a _____ point that satisfies the equation should be found to verify that the line is graphed correctly.

- For real numbers a and b, the graph of $y = b$ is a _____ line and $x = a$ is a _____ line.

The Slope-Intercept Form: $y = mx + b$

- For a line, the ratio of _____ to _____ is called the slope of the line.

- Plot the rise and run between points on the line in the graph below:

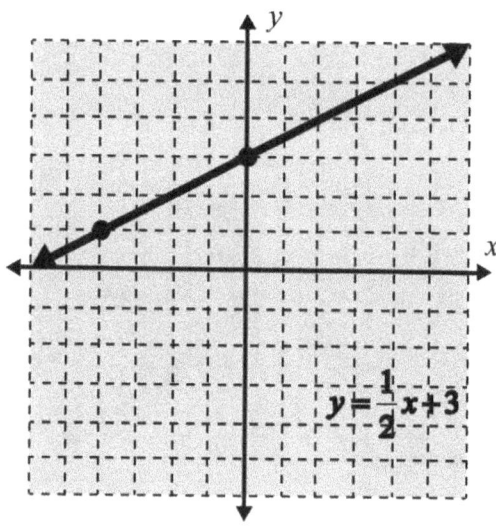

- The concept of slope also relates to situations that involve _____ of _____.

- _____ Formula = $m = \dfrac{rise}{run} = \dfrac{y_2 - y_1}{x_2 - x_1}$

- The variable that represents slope is _____.

- The slope is the same even if the order of the points is _____.

- The coordinates must be _____ in the same order in both the _____ and the _____.

- Lines with positive slope go up (_____) as we move along the line from _____ to _____.

- o Lines with negative slope go down(_____) as we move along the line from _____ to _____.

- o If two points have the same _____, such as (2, 8) and (7,8), then the line through these two points will be _____.

- o If two points have the same _____, such as (2, 3) and (2,−2), then the line through these two points will be _____.

- o The following two general statements are true for horizontal and vertical lines:

    1. For horizontal lines (of the form $y = b$), the slope is _____.

    2. For vertical lines (of the form $x = a$), the slope is _____.

- o $y=mx+b$ is called the _____-_____ form for the equation of a line, where m is the _____ and $(0, b)$ is the _____.

Notes

Notes